Project Management
Starter Kit
for Developers

Vivek S Gupta

COPYRIGHT

DEDICATION

This book is dedicated to Dr. Kamala Bhandari who was from Bikaner, Rajasthan in India.

Dr. Bhandari always believed in me. She helped me understand why it is desirable to pair gentle words with forward looking actions and good intentions.

PREFACE

The goal for the book is to help developers successfully manage projects. It can also help development managers or scrum masters manage projects. Individuals who are new to the project management field can also learn the ropes. I hope to explain how you can have successful projects along with a strong foundation in the project management field.

I have seen so many talented individuals who are not always successful up to their potential. One of the reasons is that they do not understand the fundamentals of what makes a project team click. I hope to provide insight on how to better understand your project team and help your project team also better understand the project.

The book is written primarily for software development projects. The goal is to give you an insight of various aspects of project management. You will need to review additional material to gain more in depth knowledge on numerous topics like scope or schedule management.

This book aims to help you better manage your projects, evaluate the state of your project at any given time and determine what course correction to undertake in case you need to get your project back on track.

TOP QUALITIES TO BE A SUCCESSFUL PROJECT MANAGER

Be a leader
You should be able to motivate yourself irrespective of external factors. You should be a high energy person who can pull the team through tough phases. You need to lead by example and set the bar high for yourself. You should hold everyone accountable including your-self in a fair and consistent manner.

Ability to influence others
You should build credibility with the team. You should have credibility to steer decisions towards what is best for your project.

Have excellent communication skills

Active listening
It is very important that you truly listen to your team. It is critical that you let them complete their thoughts without interrupting. Where practical, you should echo back what you heard to validate that you correctly captured the essence of what you heard.

Effective written and oral capability
You should have both vocal and written skills. You should speak effectively at meetings. Your presentations should be crisp and to the point.

Enjoy problem solving

You should enjoy identifying problems and solving them. You should not passively request others to problem solve. At a minimum, you should facilitate sessions where you clearly state the problem, bring appropriate subject matter experts together and guide them to a practical solution which suits your overall project goal. Team members should know that you are the go-to person when they have problems.

Are a natural people person

You must have a natural disposition to work with people. You should look forward to individual and group interactions and thrive on it.

Be a team player

You should be secure enough to truly share success with others. You should be able to effectively work with team members with different cultures, backgrounds and abilities. You should be open to different opinions and approaches even if you do not initially agree to that view point. You should strive to see that the interests of all stakeholders are taken into consideration for decision making.

Be open to working outside your comfort zone

You should not be hesitant to learn new tools or skills to be a better project manager who is better suited to your organization needs.

Lead from the front
You need to show the way to your team. You should have the confidence to have both an emotional as well as a factual conversation.

Be open to admitting your mistakes
You should always be the first to own up to your mistakes and propose how you plan to take corrective action. This will create an open culture where it is safe to admit mistakes.

Be flexible
There are many ways to achieve goals. You should be open to choosing the best possible solution irrespective of who proposes the solution.

Learn when to back off
You also need to figure out on when to back off and not hinder progress. It is important to not always push people on every item even when it is minor. Your team should know that you trust them, especially when they are proven performers.

Build Trust
Try to build a culture where team members trust each other and help others succeed. This is the most essential ingredient for a successful project.

Evolve

You have to start with yourself. It is not be possible to achieve everything from day one. You need to continuously evolve. Patiently aim to be a better version of yourself today when compared to yesterday.

The sky is the limit if you and your team truly follow these basics:

1. Treat your team members the way you expect to be treated
2. Give credit where due
3. Be respectful at all times
4. Treat everyone with courtesy
5. Readily admit your mistakes
6. Never humiliate anyone
7. Share the limelight
8. Ask for help from the subject matter experts
9. Actively support innovation
10. Actively support diversity of thought
11. Have fun!

APPROACH

First get the fundamentals "right"
I recommend that you first understand the numerous aspects of project management. Understand how all these aspects interplay with each other. Get a feel on how you can harness great synergy within your project when you leverage the various aspects correctly.

Become a "true" subject matter expert in your field
Aim to be an expert in the aspects once you grasp the fundamentals. You need to pay attention to every aspect for a smooth running project. The best way you can differentiate yourself from other project managers is to know when to stress more on which aspect and for how long.

Customize for your team
I am not a big fan of rigid processes and templates. The best process is what fits your project team and your organization culture. There are numerous ways to solve the same problem. I recommend that you customize the process and templates to best suit your organization and project team.

Have only value add processes
I am all for process when it adds value. Do not have a process if it has little value. Avoid your team wasting time updating templates if there is little or no value to doing so.

25 ASPECTS OF THE PROJECT MANAGEMENT DASHBOARD

You need to manage your project on multiple fronts. You should pay appropriate attention to the various aspects of your project.

Imagine that you have a project management dashboard which you will use to manage your project. It is critical to not overlook any part of your dashboard. It is equally critical to have self-awareness on where to focus and when to focus depending on your project state. You can view yourself as a pilot in a cockpit where every instrument gauge is important. You need to understand every instrument, when to refer which instrument and the degree of action to take depending on various situations.

Your project dashboard has the following 25 characteristics:

1. Project Goals
2. Project Champion
3. Project Stakeholders
4. Project Team
5. Know Your Customers
6. Meetings
7. Share Status Reports
8. Financial Management
9. Scope Management
10. Change Management
11. Schedule Management
12. Critical Path Management
13. Resource Management
14. Issue Management

15. Risk Management
16. Quality Management
17. Priority Management
18. Show Stopper Management
19. Interdependency Management
20. Multiple Projects Management
21. Conduct Trade-Off Assessment
22. Marketing Your Project
23. Support Innovation
24. Talent Management
25. Partnership

SECTION I: WHY

1) PROJECT GOALS

A) What is your Project Goal?

The project goal is the objective of your project. It is the "success criteria" which your project needs to achieve. Your goal should be measurable and concise. It is the elevator pitch of what your project stands for.

B) Why is important to have a project goal?

The project goal provides meaning to your project. It helps you defend why your project is required. It helps your project team understand why they should contribute to make your project a success.

C) Who should you involve in creating your project goal?

The Project Champion is the most critical resource in defining the project goal. Certain key stakeholders who expect certain benefits or are impacted by the project should also be involved in creating the project goal.

D) Format for Project Goal

The goal of the project is to provide *"service or key features"* to achieve *"benefit"* by *"time frame"*.

E) Example for Project Goal

The goal of the project is to create a mobile interface for the financial application so that customers can see their account balance and transactions information. Customers should also have an ability to create dynamic reports. The expected capability is expected to be piloted by Q2 of 2017 and be available for all customers by Q4 of 2017.

SECTION II: WHO

2) PROJECT CHAMPION

A) Who is the Project Champion?

The Project Champion is the person in your organization who wants your project to succeed. He or she is most vested in your project success. Your project champion is usually a Senior Executive or a Product Manager who is able to convince your organization on why your project needs to be initiated.

B) Why do you need a Project Champion?

Every project is expected to run into obstacles and challenges. Most projects compete with other projects for various types of resources (funds, hardware, people, etc.). Having a Champion who has clout in the organization in your corner will make the path to success comparatively smoother.

C) What do you need to do if you do not have a Project Champion?

The short answer is to identify a Champion before it is too late. Projects which do not have clear Champions are usually not successful and are unpleasant experiences for the project team. Projects with numerous stakeholders with no clear Champion are usually stressful to the project team as there is no single person who has the authority to make tough decisions when there are competing interests on the same project.

You should seriously consider stopping the project if no one really cares if your project is implemented or not. A project which does not waste your resources is a success vs. a project which is successfully implemented, has consumed resources and has little to no use.

D) What do you need to do if your Project Champion has no time for your project?
At a minimum, you need to have an understanding with your Project Champion that you will reach out to him or her when you need support. You should continue to send regular status reports even if the project is doing well so that the Project Champion is aware that your project is on the path to success.

3) PROJECT STAKEHOLDERS

A) Who are project stakeholders?

A stakeholder is a person or team who is impacted by your project. The impact can be positive (example: new features), neutral (example: hardware upgrade with no change to functionality) or negative (example: access is removed for an existing feature).

B) Why is it important to identify project stakeholders?

Every stakeholder can positively or negatively impact your project. Having stakeholders aligned with you will ensure that you have support when remove obstacles for your project team.

C) How to handle a situation when there are competing stakeholders?

The Project Manager needs to keep all lines of communication open with stakeholders, especially when there are competing stakeholders. The Project Manager is expected to align all stakeholders towards the common goal. The Project Manager should truly try to understand the stakeholders perspective if they have competing priorities. It is best to resolve these differences early and in a professional fact based manner. Competing items should be negotiated fairly to ensure stakeholders remain supportive of the project.

D) Why is it important to actively manage project stakeholders?

The Project Manager is responsible to ensure that no stakeholder is surprised. The Project Manager should actively listen to stakeholders. The Project Manager should either revise project goals or align stakeholder expectations with project goals. Ignoring issues with stakeholders will result in negativity and unnecessary noise on the project. The Project Manager should enlist the Project Champion's help in case he/she does not have enough clout in the organization to get stakeholders to align with the project goals. The project goals cannot be changed without explicit buy-in from the Project Champion.

4) PROJECT TEAM

A) Who are your project team members?

Project team members are the ones who help the project achieve its goals.

B) Why is important to clearly state your project team members?

Your project team members have deliverables for your project. It is important for you to have their true buy-in that they will contribute to the project in a timely manner.

C) Why is important to have a fabulous relationship with your project team members?

The project team can make or break your project. You will naturally have a good relationship with team members when there is clarity on goals, issues are proactively resolved and team members provide true buy-in on deliverable milestones. You need to be fair to every team member and not be biased towards any role or level. A fair environment where success is shared ensures that the team works well together and with you.

D) What do you have to do if your project team members do not want to contribute to your project?

Escalations don't work! Escalations burn bridges and bring negativity to the team's culture. You need to identify the root cause on why someone does not want to provide the deliverable or not provide it in the expected timeframe. You should work with the team member to remove the roadblocks. If needed, you and the affected team member should "together" leverage the team member's management team to remove the roadblocks. Speaking with one voice to achieve the same goal will go a long way to resolve the situation. Recommend options with pros and cons when you meet with the management team. There is a probability that you may need to include the Project Champion and stakeholders in case there is major negative impact to the project.

5) KNOW YOUR CUSTOMERS

A) Who are your Customers

You need to identify both internal and external customers. Your customers can also be your project stakeholders.

B) Why do you need to identify your Customer expectations

It is important to ensure your project goals are aligned with customer expectations. At the start of the project, identify the metrics which will determine if you met customer expectations or not. Collecting these metrics post implementation will ensure that there is fact based conversations post implementation on whether the project was successful or not.

C) How to handle missed Customer expectations

The key is to be honest and upfront. Do not surprise your customers with bad news. Explain the reasoning on why you missed the goal. Solicit feedback on viable options which can help you regain lost ground.

SECTION III: HOW

6) MEETINGS

A) Do you really need a meeting?

Do not schedule a meeting if you don't need one. Try your best to eliminate as many meetings as you can. Try to not have meetings with large number of people as they will not be productive. An offline conversation is usually more productive than a formal meeting.

B) Messaging for your audience

You should tailor your agenda to your audience and the agenda topics. At a minimum, you have three types of meetings

1) Stakeholder meetings

2) Project meetings

3) Ad-hoc meetings which are for specific topics

C) Steps to follow prior to the meeting

Schedule the meeting when participants are available. When you send the invite, clearly state the following

1) Meeting objective

2) Agenda Topics - each agenda topic should be time boxed

3) Required attendees

4) Optional attendees

5) Meeting location

6) Meeting start and end times

7) Dial-in information where applicable

D) Steps to follow during the meeting
Take roll call and state the objective of the meeting. Request your audience to ask clarifications if the meeting objective is not clear. It is important that the participants agree on why they are meeting before you go over each agenda item.

Go over the agenda items and solicit feedback if new agenda items need to be added to meet the meeting objective. This will ensure that participants do not derail your meeting as they know that their topics are going to be discussed.

Guide your team to stay on topic. If new topics are raised, you can put them on the parking lot for future meetings. You can also make a judgment call on letting new topics be discussed and let topics go over their allocated time. This will lead to additional meetings be scheduled at a later time.

Try to get a sense of how your project team gets to decisions and allocate sufficient time for topics. Forcing a decision without true buy-in will only lead to issues and confusion later.

At the same time, do not let a topic drag on if there is no need. Project team members appreciate if a meeting ends early as long as all the decisions are made.

E) Steps to follow after the meeting

Figure a way to get anonymous feedback from participants on whether the meeting was successful or not. This will go a long way in ensuring that you are not wasting time.

Within 24 hours of the meeting, send the following to invitee list along with any impacted stakeholders

1) Attendee List

2) Key Points

3) Action items with owner names and agreed upon due dates

F) Format for Meeting Invite

Meeting objective:
Required Attendees:
Optional Attendees:
Chairperson:
Meeting Location:
Meeting Date:
Meeting Start Time:
Meeting End Time:
Dial-In:
Agenda Topics *<Topic 1, presenter(s), XX minutes>* *<Topic 2, presenter(s), XX minutes>* *<Topic 3, presenter(s), XX minutes>* *<Topic 4, presenter(s), XX minutes>* *<Next Steps, chairperson, XX minutes>*

G) Format for Meeting Minutes

Meeting objective:
Required Attendees who attended:
Required Attendees who could not attend:
Chairperson:
Meeting Date:
Key points: *<Key point 1>* *<Key point 2>*
Action items: *<Action item 1, Owner Name, Due Date>* *<Action item 2, Owner Name, Due Date>* *<Action item 3, Owner Name, Due Date>* *<Action item 4, Owner Name, Due Date>*

7) SHARE STATUS REPORTS

A) Communicate based on your audience and their interests

At a minimum, you should create status reports for three audiences 1) Project team 2) Stakeholders 3) Senior Management. The level of reporting and what to communicate should be tailored to your audience. As an example, Senior Management will not be interested in minor issues and will prefer to get the big picture. You should also be explicit if you are requesting Senior Management for support on certain issues to ensure project success. Ideally, your project rating should be the same for all status reports. Your project's future is dependent on how you convey your status. You should not surprise your audience by flip flopping between your project ratings. Your audience should have confidence that you are truly all over the issues and are managing the project with a steady hand.

B) Project Status Reports

Status reports for your project should be specifically tailored to your project team members. Every key issue and risk should be mentioned. The messaging should be at a more detailed level. Ideally, these status reports are one to two pages long.

C) Stakeholder Status Reports

Stakeholder status reports should be specifically written on what the stakeholder is looking for. The messaging should be focused on the needs of the stakeholder so that the status report is relevant to the stakeholder. Stakeholder status reports should fit within one page.

D) Senior Management Status Reports

Senior Management status reports should convey the big picture. Senior Management status reports should also fit within one page. It is very important to not gloss over critical issues and risks in this status report. At the same time, you should not unnecessarily cause panic on your project if there is no sense of doom. Bottom line, you have to be very accurate and concise in this status report.

E) Format for Status Reports

Project Name: *<Name>*
Project Goal: *<Goal>*
Project Status As of: *<Date>*
Project Rating: *<Red, Amber, Green>* trending *<Red, Amber, Green>*
Key Message: <2-3 Sentences on why does the project have a certain rating and high level steps on how to keep/get to a Green status>
Accomplishments: *<Accomplishments 1>* <Accomplishments 2>
Upcoming Activities: *<Upcoming activity 1>* *<Upcoming activity 2>*
Milestones: *<Milestone Name 1, Rating (Green, Amber, Red, Not Started, Completed), Start Date, Due Date, Owner>* *<Milestone Name 2, Rating (Green, Amber, Red, Not Started, Completed), Start Date, Due Date, Owner>*
Issues: *<Issue Description 1, Rating (Green, Amber, Red, Resolved), Impact, Resolution Steps, Original Due Date, Revised Due Date, Owner>* *<Issue Description 2, Rating (Green, Amber, Red, Resolved), Impact, Resolution Steps, Original Due Date, Revised Due Date, Owner>*
Risks: *<Risk Description 1, Rating (Green, Amber, Red, Mitigated, Accepted), Potential Impact, Mitigation Steps, Original Due Date, Revised Due Date, Owner>* *<Risk Description 2, Rating (Green, Amber, Red, Mitigated, Accepted), Potential Impact, Mitigation Steps, Original Due Date, Revised Due Date, Owner>*

8) FINANCIAL MANAGEMENT

A) What is my budget?
Your project is the funds allocated to implement your project goals. Your project budget may be allocated on an annual basis

B) How do I manage by budget?
It is always important to remain within budget. You should first add your actual costs to your forecasted cost. You manage to the budget by comparing this total to the allocated budget.

C) What to do if you predict you will be over budget?
You may need to adjust other project characteristics to remain within budget. As an example, you can reduce cost by reducing scope by not delivering the least important features or the feature which is resource intensive and has comparatively lower priority. You can also delay features to the future in case you are managing to an annual budget. You need to, of course, ensure stakeholder and executive buy-in when you propose various options on how to remain within budget

D) What to do if you predict you will be under budget?
You can accelerate delivery of features as long as your team remains efficient. You can also choose to invest in improving

the quality of the product. You can also return funds back. Again, ensure stakeholder and executive buy-in when you propose various options on what to do with the additional funds.

9) SCOPE MANAGEMENT

A) What is scope?

Scope is the list of features your project team has to deliver to achieve your project goals. The attributes associated with each feature are:

1) Feature Number
2) Feature Name
3) Feature Description
4) Expected impact to the customer
5) Why is the feature required
6) What is the cost of the feature
7) What is the priority (Urgent, High, Medium, Low) of the feature
8) What is complexity (Critical, High, Medium, Low) of the feature
9) What feature(s) need to be created before this feature is delivered
10) What feature(s) are dependent on this feature being delivered
11) What is the impact to the customer if the feature is not delivered or has major defects

B) What is scope creep? How do I avoid it?

Scope creep or gold plating is when the project team or stakeholders try to increase scope by additional features which do not have value to the customer based on the cost incurred to deliver these features

C) How do I manage scope?

You should maintain a list of features currently in scope. You should also have a healthy feature backlog in case in-scope features are cancelled and additional scope can be added.

You should explicitly call out interdependencies with partner teams and ensure they are in-sync with what is expected and when.

You should explicitly call out "out of scope" items at the very start of the project to ensure everyone remains on the same page

D) How do I avoid gold plating?

You should share the cost of each feature. This will help the team and stakeholders to not request that unnecessary features are built as they will realize the additional cost to the project

10) CHANGE MANAGEMENT

A) What is change control?
Change control is managing any change to your project in an effective, fair and consistent manner.

B) Why is change control necessary?
It is important to capture changes to existing scope or schedule along with their impact on existing commitments. Changes must be approved before the team agrees to the change

C) Who is part of the Change Control Board?
Sponsors and Stakeholders are part of the Change Control Board.

D) What is change control process?
Step1: Anyone can submit the change.
State what is the change, why is it required and when is the change expected to be implemented. It is also important to explain the impact if the change is rejected

Step 2: Change Control Board approves (or rejects) whether the change can be evaluated.

Step 3: The team who will work on the change evaluates the change

Determine effort and cost estimate. Determine impact on existing commitments.

Step 4: Change Control Board approves (or rejects) the change based on the evaluation

11) SCHEDULE MANAGEMENT

A) What is a project plan?

A project plan is a list of tasks. The tasks roll up to major and minor milestones to show project progress. Complex tasks can also be broken down to sub tasks. The task should usually lead up to a deliverable which adds value in achieving the project goals.

The tasks and milestones are usually dependent on other tasks and milestones. They have definite start and completion dates along with percentage of completion and resource names. A resource can be partially allocated to a task.

B) Do you need to create a project plan in Microsoft Project?

No. You can choose any tool to create a project plan.

C) Why do you need a project plan?

It is important for your team members to contribute as well as understand the deliverables as well as the timelines

D) When do you share your project plan?

You start sharing the first draft of the project plan as early as possible. This is to ensure that your team as well as stakeholders can start influencing the schedule from the very beginning

E) Who contributes to the project plan?

Everyone on the project team contributes to the project plan. You will get true ownership of tasks and deliverables if your team members truly agree to the plan

F) What to do if deliverables cannot be created on time?

You should try to see how delaying the deliverable does not delay the overall project. As an example, try to determine if a solid draft of the deliverable can be created as per the original date. This can allow the subsequent deliverables to continue without too much of an adverse impact

G) What to do if I have an unrealistic timeline?

This is something which you cannot let fester. Your team will be thoroughly demoralized if they do not believe in project success. You may need to adjust other project characteristics to remain within schedule. As an example, you can reduce project duration by reducing scope by not delivering the least important features or the feature which is resource intensive and has comparatively lower priority. You can also delay features to the future release. If you have funds and your project remains efficient, you can also choose to get additional resources to reduce the timeline. You need to, of course, ensure stakeholder and executive buy-in when you propose various options on how to remain within schedule

12) CRITICAL PATH MANAGEMENT

A) What is your project's Critical Path?

A Critical Path is all the milestones (or tasks) which must be completed in a certain order to achieve your project goals. The critical path is expected to be completed in the shortest time possible

B) Why is it important to regularly share and discuss Critical Path

It is very easy for the team to get distracted by numerous tasks which take them away from achieving project goals. Some team members have the tendency to lean towards gold plating by delaying critical path tasks. Bringing up the critical path is one way to avoid gold plating by having a factual conversation

13) RESOURCE MANAGEMENT

A) Who are the resources on your project?

Anyone who contributes to your project is a resource

B) How to handle a situation when you have insufficient resources?

You need to figure out on how to best handle the situation. You need to assess if you can even request resources depending on the political climate of the organization. You need to have a solid business on why you need additional resources. Your case should explicitly list how many resources are requested along with the duration and cost. You need to also communicate how the project will be impacted if the resource request is approved or denied. You should be realistic in the business case and state facts. Having a fair business case with options with options and contingency is expected by the decision makers.

C) How to handle a situation when you have idle resources?

It is hopefully a rare occurrence to have idle resources in today's world. Idle resources can work on items which are planned on future deliverables. Capable resources who idle are demotivated and lead to project failure as the team performance is negatively impacted. Idle resources can also be used to improve the quality of existing deliverables. It is also a good idea to temporarily loan idle resources to another project

D) How to track resources?

Keep it simple. Have a chart of resources by month

Example:

Resource type

Cost

Location

Availability

14) ISSUE MANAGEMENT

A) What is an issue?
An issue is an event or activity which will adversely impact the project.

B) Why is it important to resolve issues quickly?
Unresolved issues which fester unnecessarily too long reduce the confidence in the project success. Unresolved critical issues will mostly cause project delays and/or project failure. Issues not resolved will usually turn critical if no one pays attention to them.

C) When do you discuss issues?
At a minimum, the status on issues should be discussed at project status meetings. There should be separate discussions to resolve critical issues in a timely manner.

D) Issue Tracking Format

Issue No.	1
Issue Description	2-3 sentences on what is the issue
Rating	*Green, Amber, Red, Resolved*
Impact	2-3 sentences on the impact of the issue
Resolution Steps	2-3 sentences on what resolved the issue. This will, of course, be completed after the issue is resolved
Opened by	The name of the person who raised the issue
Opened on	The date when the issue was raised
Original Due Date	The initial target date to resolve the issue
Revised Due Date	The revised target date to resolve the issue in case the initial target date will be missed
Owner	The person who is responsible to resolve the issue

15) RISK MANAGEMENT

A) What is a risk?
A risk is an event or activity which has some probability of adversely impacting the project.

B) What are the two actions you can take on a risk?
You can choose to mitigate the risk. Alternatively, you can choose to accept the risk.

C) Why is it important to take action on risks quickly?
Unresolved risks which fester unnecessarily too long reduce the confidence in the project success. Unresolved critical risks will mostly turn into issues which can cause project delays and/or project failure.

D) When do you discuss risks?
At a minimum, the status on risks should be discussed at project status meetings. There should be separate discussions to take action on critical risks in a timely manner.

E) Risk Tracking Format

Risk No.	1
Risk Description	2-3 sentences on what is the risk
Rating	*Green, Amber, Red, Accepted, Mitigated*
Probable Impact	2-3 sentences on the impact of the risk
Mitigation Steps	2-3 sentences on what mitigate the risk. This will, of course, be completed after the risk is mitigated. You can use this column to explain the reasoning on why the risk was accepted
Opened by	The name of the person who raised the risk
Opened on	The date when the risk was raised
Original Due Date	The initial target date to disposition the risk
Revised Due Date	The revised target date to disposition the risk in case the initial target date will be missed
Owner	The person who is responsible to disposition the risk

16) QUALITY MANAGEMENT

A) What is quality?
Quality is providing the best possible experience to the customer

B) Is it acceptable to have issues after implementation?
One should always aspire to not have issues in production. It is absolutely not acceptable to have major issues after implementation. One should always set expectations with stakeholders in case there will be quality issues after implementation due to trade-offs during development. The stakeholders are the ones who should make the final decision during trade-off discussions as they are the ones who have to face the consequences of the decisions

C) What do I manage quality on my project?
Defects are to be tracked throughout the project life cycle and only during testing. All deliverables are to be reviewed for quality. Defects are expected to be prioritized, tracked and dispositioned. Higher priority defects should be resolved vs. deferred

17) PRIORITY MANAGEMENT

A) What is priority? What is Priority Management?

Priority signifies the importance of a task, issue, risk, feature or a milestone. Priority Management is relentless focus on the most critical items first.

B) Why do you need to manage priorities?

The nature of our work is such that there are numerous distractions throughout the day. Our work is never ending. There is no pride in taking unfinished business home because you mismanaged your work day. Working based on priorities will ensure that you drive a positive culture in your team where the focus remains on doing what's best for your Clients in the fastest possible way.

C) How can you better manage your priorities?

You need to carve out 15 minutes at the end of the day to determine the priorities of the subsequent day. This will ensure that you hit the ground running when you start the next day. You should also regularly do a self-evaluation to determine if you are working as per pre-determined priorities or are you letting others drive your day. It is important to make course corrections in case you determine that more than 80 percent of your pre-determined priorities are not accomplished. It is highly recommended to plan some portion of your day to deal with the unexpected as well on innovation. This will keep your mind fresh and you will have more control of your day.

18) SHOW STOPPER MANAGEMENT

A) What is a show stopper?
A show stopper is any anything which can cause your project to fail. Show stoppers are mostly issues which must be resolved before a project goes live. Show stoppers are usually the most important issues which you need to drive constantly towards resolution

B) Why do you need to actively manage show stoppers?
Unexpected show stoppers cause project failure, demoralize teams and ruin careers. You need to actively manage show stoppers for yourself, your team and your organization. You are responsible for project success

C) How do you avoid show stoppers in the first place?
The obvious one is to actively manage issues. You need to regularly probe to identify new issues. You create a culture where issue identification is celebrated as much as issue resolution.

D) How do you solve show stoppers?
You should be open to temporary workarounds to mitigate show stoppers. It is important to reprioritize when you run into show stoppers. This will free up the appropriate subject matter expert to remove quality time to resolve the show stopper. Worst case scenario, you may need to renegotiate items like scope and schedule to resolve show stoppers

19) INTERDEPENDENCY MANAGEMENT

A) What is interdependency within a project?
Each project activity should lead to a value added deliverable. Each activity should have an owner, start date and end date. With the exception of "project start", each activity should have a predecessor which should trigger the activity. The activity is usually marked as complete when the outcome or deliverable is achieved. The deliverable usually triggers the start of another activity with the exception of "project end". Hence, various activities or tasks are interdependent on each other. Similarly, activities and deliverables are also interdependent

B) How to manage deliverables inter-dependent on each other?
It is good practice to have a single owner for each activity or deliverable. An activity or task should not be longer than a five day duration. There should be true buy-in from the task owner on the task and deliverable attributes. The team should explicitly discuss status on critical path as well as interdependent items on a regular basis

C) How to provide clarity on interdependent tasks and deliverables?
How to provide clarity on interdependent tasks and deliverables?
You should solicit feedback from the project team on what works best for them. It is not a good idea to force a format on

the team if they do not understand the information. It is best to create something simple to understand, change and communicate

20) MULTIPLE PROJECTS MANAGEMENT

A) Why is it important to identify and manage interdependencies between projects?

A program consists of a number of interdependent projects. It is important to understand the interdependencies between projects so that one project does not adversely impact another project. This ensures program success

B) Where to start?

You need to have a dashboard for each project

C) What else do you need to explicitly manage?

You need to identify all the touch points between the two projects. Each touch point should explicitly state what deliverable is required, who will provide it along with who will consume it. Moreover, there should be clarity on the impact if the deliverable is delayed. It is important to convey if the deliverable is on the critical path for the consumer

D) How to avoid surprises?

It is important to regularly discuss the status for deliverables which are across projects. The issues for these deliverables are usually given a higher priority as the impact is across a program

21) CONDUCT TRADE-OFF ASSESSMENT

A) What is a trade-off assessment?

There will be many times when you need to decide between key project drivers like scope and schedule. Trade-off assessment is to determine how you can achieve your project goals with certain compromises

B) Why is it important to conduct a trade-off assessment?

Most projects run into situations which require a fresh look at what is practical to achieve the project goals. Frequent assessments ensure that your project is always moving towards a delivery as per stakeholder expectations. These assessments also ensure that you don't surprise your project stakeholders

C) How does one share the trade-off assessment?

It is always important to present multiple options where the advantages and disadvantages of each option are clearly stated. This will ensure that the key decision makers can make the best possible decision on the project

22) MARKETING YOUR PROJECT

A) How to market your project?
You need to ensure that there is clarity on your project's progress. It is equally important to share what value is provided to the Client along with the adverse impact to the Client if the project is not completed as expected

B) Should you support canceling your project?
Absolutely yes, especially if it means higher priority projects need to succeed before your project. Your organization goals are more important than your project

C) Why is it important to market your project?
Every organization has competing priorities. Your project will compete with other projects for resources in order to succeed. You need to always do what is best for your organization in a factual manner and without emotion. Marketing your project based on facts will ensure that you actively support correct decision making within the organization

23) SUPPORT INNOVATION

A) What is innovation in the context of a project?
Innovation is solving problems creatively and without traditional approaches

B) Why is it important to have a culture of innovation?
Every project is expected to run into the unexpected. Embracing creativity will help the team resolve contentious issues without adversely impacting the project

C) Why is it important to learn to fail quickly?
Every innovation attempt does not lead to success. Having an ability to fail quickly will help the project team to always look for ways to innovate

24) TALENT MANAGEMENT

A) Why do you need to promote talent?
First and foremost, this is the right thing to do! Resources will want to work on your team when they know that you truly are vested in their career.

B) Why do you need to help promote individuals?
There is true joy in helping others succeed. Supporting others will also help you in building long partnership which is not transactional. These partnerships will turn into mutually beneficial and help you, your teams and the organization.

C) Why do you need to recommend individuals and teams for awards?
It is important to share the limelight and glory. Team members will flock to your team when they know that you are sincere in recommending awards. Your Management will also be appreciative if you sponsor well deserved awards.

25) PARTNERSHIP

A) What is partnership?
Partnership is having rock solid symbiotic long term relationships with individuals and teams

B) Why do you need partners to succeed?
Teams accomplish more than individuals. No one wants to help individuals or teams who are selfish or difficult to work with. There is joy in getting along with others. You and your teams will find it easier to remove roadblocks when you have others helping you on the way. The best battle is won when you don't need to fight!

C) How do you cultivate partnership?
Success comes naturally to those who help others without expecting something in return. A genuine relationship is way better than a transactional relationship. Focus on building genuine long term relationships. Always volunteer to support others and root for the success of others

ABOUT THE AUTHOR

Vivek Shivratan Gupta has a bachelor's degree in Computer Engineering from Maharashtra Institute of Technology in Pune, India. He also completed his Masters in Science in Software Engineering (Major: Project Management) from DePaul University in Chicago, United States. He has also a UNIX/Networking/C++ certificate from Illinois Institute of Technology, Chicago. Vivek is also a Six Sigma Green Belt. Vivek has the Project Management Professional Certification from the Project Management Institute in Pennsylvania, United States. He is a Certified Scrum Master from Scrum Alliance in Indianapolis, United States.

Vivek started his career as a software developer. He has played various roles some of which are: Project Manager, Program Manager, Agile Transformation Lead, Director of Application Development and Director of Project Management.

Vivek had the pleasure to live and work in three wonderful countries, viz. India, England and United States of America.

On a personal front, Vivek loves spending time with his family, reading books on various topics and travelling to different countries to experience various cultures. He also loves running half marathons.

www.ingramcontent.com/pod-product-compliance
Lightning Source LLC
Chambersburg PA
CBHW050523210326
41520CB00012B/2411